YOUR KNOWLEDGE HAS VALUE

- We will publish your bachelor's and master's thesis, essays and papers

- Your own eBook and book - sold worldwide in all relevant shops

- Earn money with each sale

Upload your text at www.GRIN.com and publish for free

Bibliographic information published by the German National Library:

The German National Library lists this publication in the National Bibliography;
detailed bibliographic data are available on the Internet at http://dnb.dnb.de .

Imprint:

Copyright © 2016 GRIN Verlag, Open Publishing GmbH
Print and binding: Books on Demand GmbH, Norderstedt Germany
ISBN: 9783668443945

This book at GRIN:

http://www.grin.com/en/e-book/364626/from-a-sustainable-development-perspec-
tive-is-nuclear-energy-a-curse-or

Ridwan Bello

From a Sustainable Development perspective, is nuclear energy a curse or a blessing?

GRIN Publishing

From a Sustainable Development perspective, is nuclear energy a curse or a blessing?

Ridwan Bello

Table of Contents

I. Introduction

Sustainable development refers to "development that meets the needs of current generations without compromising the abilities of future generations to meet their own needs" (Bruntland Commission, 1987). Critical to the attainment of this development is the availability and use of energy. According to NEA (2000), the way energy is produced and used plays an essential role in all three dimensions (economic, social and environmental) of sustainable development. To those who have it, modern energy unlocks access to improved healthcare, improved education, improved economic opportunities and even longer life; to those who do not, it is a major constraint on their social and economic development (Africa Energy Outlook, 2014).

The world's energy mix is heavily dependent on fossil fuels, which accounted for 81.4% of the world's total energy consumption in 2014 (IEA, 2015). These fossil fuels are responsible for producing a substantial amount of the world's atmospheric co_2 gas – the chief culprits causing climate change. With the implications of climate change for humanity in mind, it has become obvious that energy without further pollution is a 'must-go path' (Jeong et al., 2010). Renewable energy and nuclear energy, each with merits and demerits, have been proposed as viable alternatives (Verbruggen, 2008; Mbarek et al., 2015).

The aim of this essay is to critically assess nuclear energy and to argue that the technology is more of a blessing than a curse, from a sustainable development perspective[1]. To justify this position, this essay shall examine nuclear energy in France and Japan as case studies. These two countries – one in which nuclear energy has been put to good use, and another where it has been calamitous – were chosen in order to present a balanced argument.

II.France

The beginning of France's dependence on nuclear energy can be traced back to the oil crisis of 1973 (Petit, 2013). That year, following an embargo placed on oil production by the Organization of Petroleum Exporting Countries, and the consequent skyrocketing in oil prices, the then Prime Minister of France, Pierre Messmer launched what is today known as the Messmer Plan. The aim was to reduce France's reliance on imported oil – the main energy source at the time – and switch to nuclear energy with which the country had prior experience during World War II (EDF, nd; Teräväinen et al., 2011). Consequently, by 1989, the country had constructed a total of 56 nuclear reactors, and currently operates 58 reactors (WNA, 2015). During the time, the share of nuclear energy in the country's domestic electricity supply rose from 4% in 1970 to 24% in 1980, and 78% by 2006 (IEA, 2006).

An examination of France's pro-nuclear stance across the three domains of sustainable development throws up interesting issues. From an economic perspective, the reliance on nuclear power has been rewarding. Unlike in the years preceding the 1973 oil crisis, the country has apparently acquired greater control of her energy generation and supply in recent years. This is immensely important, considering that such security is a vital requirement for economic growth and development. In addition, the country is the largest net electricity exporter in the world (Teräväinen et al., 2011), raking in over €3 billion annually from electricity export (WNA, 2015). For a country with no vast coal, oil nor gas deposits, and whose total natural resource rents accounted for just 0.1% of GDP between 2011 to 2015 (World Bank, 2016), such income from electricity export is surely a fortune.

---

[1] In the context of SD, "Blessing" is taken to mean aggregate positive social, economic and environmental impacts outweigh negative impacts. "Curse" is the reverse.

Moreover, as a result of nuclear dependence, electricity prices in France are among the lowest in the world (Brook et al., 2014), and electricity users in France pay, for the same unit of electricity, about half as much as their counterparts in "anti-nuclear" countries like Italy and Germany (Statistica, 2015). High energy prices have been associated with high rates of household energy poverty in the EU (Bouzarovski, 2014), and given that one of the Sustainable Development Goals is to provide affordable and clean energy, low electricity prices in France is certainly desirable.

In terms of environmental impacts, nuclear energy also appears to have paid off for France thus far. "The country's per-capita GHG emission is among the lowest for industrial nations worldwide and many times lower than in otherwise similar countries (such as Australia and Denmark) that have no nuclear power plants and that rely on a mix of fossil fuels and some contribution from renewables" (Brook et al., 2014). Thus, at a time when other industrialised but less nuclear-reliant countries like the UK, China and Russia are fretting over their co_2 emissions in attempts to mitigate climate change, nuclear energy has afforded France some breathing space.

Moreover, while there have been some nuclear plant mishaps, adverse environmental impacts from such have been minor. Data from the International Atomic Energy Agency show that of the 33 nuclear accidents recorded worldwide between 1952 and 2011, only two were recorded in France, neither of which caused any severe environmental impacts – the melting of a channel of fuel in the Saint Laurent reactor in 1980 caused "no release outside the nuclear site", and the accident at the Cadarache nuclear facility in 1993 which led to the "spread of contamination to an area not expected by design" was classified based on IEAE ratings as only an "incident" (IAEA, nd). Furthermore, nuclear wastes which could potentially contaminate the environment from fall-outs of radiative substances (Blowers, 1999), are being handled as safely as possible. According to WNA (2013), "most short-lived intermediate- and low-level wastes are sent for final disposal at National Radioactive Waste Management Agency's (ANDRA's)[2] surface waste repositories". In 2006, ANDRA received parliamentary approval for the construction of retrievable deep geological repository for its high-level wastes (WNA, 2013). As of 2010, construction was progressing as planned, and the country's existing high-level and medium-level radioactive wastes, as well as new wastes that will be generated over at least the next 20 years could be safely secured in possibly the world's first deep geological repository by 2025 (Butler, 2010). Surely, burying radioactive materials underground can hardly be described as "green". But neither are the GHG emissions that would have been generated had the country continued to rely on fossil fuels. As far as we know today, the latter is a more potent and definite threat to the environment.

From a social standpoint, the rarity of major nuclear plant disasters, with the potential attendant casualties, is also a massive advantage. However, in terms of being democratically decided – which according to Verbruggen (2008) is one of the sustainability criteria upon which energy technologies should be evaluated – nuclear power in France has earned some criticisms. While the generality of the French people is believed to be pro-nuclear (Stainer and Stainer, 1995, Lee and Gloaguen, 2015), there are suggestions that government decisions that led to the entrenchment of nuclear power in France right from the 1973 Messmer Plan have been somewhat authoritarian and dismissive of the voives of the feeble, but nonetheless existent, anti-nuclear movements (See Wiegman et al., 1995 and EDF, nd).

---

[2] ANDRA is the agency responsible for France's nuclear wastes management.

In more recent years, concerns over nuclear safety lingers and public opposition to nuclear energy in the country may be growing (Kessides, 2012, Petit, 2013). However, "with the government now considering a partial decommissioning of nuclear power stations by 2020 and lowering to 50% the share of nuclear energy in the production of electricity" (Petit, 2013), those anti-nuclear sentiments are finally being heard. Or so it seems.

Putting it all together, it is a fair assessment to conclude that nuclear energy has, over the past four decades, effectively powered the French economy with little adverse social and environmental consequences. For now, that is a blessing. How sustainable this blessing is into the future will hinge critically on the eventuality of finding more environmentally-innocuous methods to dispose the country's accumulating nuclear wastes (Abu-Khader, 2009), and her ability to consistently maintain safe reactors. With technological advancements towards safer reactor designs and better nuclear wastes treatment methods well underway (Hyde et al., 2008, Dautray et al., 2012), that blessing could well extend into the foreseeable future.

III.   Japan.
The case of nuclear energy in France is a pleasant one, but it is hardly typical. In some other countries, nuclear energy has not enjoyed such a smooth sail. The case of Japan especially builds the argument against nuclear energy from sustainable development perspectives.

The literature on nuclear energy in Japan is dominated by the 2011 accident at the Fukushima Daiichi Nuclear Power Plant. Understandably so. But long before Fukushima, Japan depended on imported fossil fuels for electricity generation – in 1973 for instance, 80% of her electricity was generated using imported oil, largely from the Middle East (Shibata, 1983). Shibata (1983) further notes that the oil crisis of 1973 hit Japan hard (as it did France) forcing the country to re-evaluate its domestic energy policy and diversify its energy sources to include more nuclear energy and less fossil fuels. Few years before 2011, nuclear energy had been billed to play an even more crucial role in Japan's energy future, apparently due to concerns about climate change. The WNA (2016) reports that in 2008, the Japan Atomic Energy Agency had modelled a 90% reduction in co_2 emissions by 2100 that would lead to nuclear energy contributing 60% of primary energy in 2100 (compared with 10% in 2008), renewables 10% (up from 5%) and fossil fuels 30% (down from 85%).

Then Fukushima happened. Copious details of the accident abound, but the summary is that an earthquake of 9.0 struck, triggered a tsunami that shattered the nuclear plant's defence walls and flooded it, causing explosion in three of its units (Bachev, 2014, Gallardo and Marui, 2016). There are suggestions that the occurrence and eventual scale of the disaster were attributable to poor administration and management (The Economist, 2011). Particularly, the Fukushima Nuclear Accident Independent Investigation Commission concluded that the accident "cannot be regarded as a natural disaster. It was a profoundly man-made disaster that could and should have been foreseen and prevented. And its effects could have been mitigated by a more effective human response" (The National Diet of Japan, 2012, p.9).

It is fair to say that whatever "blessing" nuclear energy had offered Japan up until 2011 took a big hit from Fukushima. An assessment of the economic, social and environmental implications of nuclear energy in Japan is likely to be skewed towards the negative by the accident.

The environmental impacts of Fukushima were considerable, at the least. According to Bachev (2014, p.5):

> *"radioactive elements were released from the plant into the atmosphere in the form of radioactive gases or radioactive particles (aerosols) dispersed into the air, a portion of which fell on the ground soil and formed residual radioactive deposits; and into the marine environment, directly in the form of liquid releases into the sea and indirectly due to fallout on the sea's surface from radioactive aerosols dispersed over the ocean."*

With vast stretches of land, nearly 1,800 km^2, contaminated with radioactive substances (The National Diet of Japan, 2012, p.38), the environmental impacts become even more profound when one considers the plethora of species of flora and fauna that would have been lost to the radiation.

From social perspectives, there were no direct mortalities from the accident (WNA, 2016), but "146,520 residents were relocated and approximately 14,000 residents (excluding plant workers) from three towns and villages where radiation doses were relatively high, were found to have been exposed to various degrees of radiation over the first four months after the accident" (The National Diet of Japan, 2012, p.38). In the initial weeks after the explosions, citizens were warned not to eat agricultural products from the region for fear of contamination, and one year later, many of the evacuated people still remained in temporary housing facilities (Schreurs, 2013). There were also temporary power shortages resulting from a complete shutdown of all the country's nuclear plants following the accident, with marked increases in electricity prices seriously impacting people's lives and the country's economy (Ming et al., 2016). Evidently, the accident had profound adverse impacts on the lives of the casualties in particular and the people of Japan in general.

In terms of economic impacts, there have been various costs estimates of the accident. Srinivasan and Rethinaraj (2013) estimates the direct economic cost "subject to unavoidable errors of measurement and possible biases" to be US$65billion. Adelman (2011) estimates the cost of decontamination alone to be around US$14billion over 30 years. According to Reuters (2011), the Japanese government projected that the total cost could be as high as US$250billion, if compensation to victims and resettlement cost were included. While these estimates may be imprecise, they at least give an impression of how huge the economic costs of the accident could be.

Taking a holistic assessment, nuclear power might have stimulated economic growth in Japan as submitted by Lee and Chiu (2011) as well as Naser (2015), but it is hard to argue that this expiates the adverse social, economic, and environmental impacts of Fukushima. Yet Fukushima is not the only nuclear accident ever recorded in Japan. According to data from IAEA, of 33 nuclear accidents recorded worldwide between 1952 and 2011, there were four other less severe accidents asides Fukushima[3]. These give the uneasy feeling that unless the technology is completely abandoned or drastic safety measures put in place, Fukushima may not be the last. Japan thus suggests that from a sustainable development viewpoint, nuclear energy should be treated with more aversion than sympathy.

---

[3] Tsuraga in 1981, Ishikawa in 1999, Tokaimura in 1999 and Ongawa in 2011 were respectively rated 2, 2, 4, 4 on the International Nuclear Event Scale.

IV. Discussion

From the two cases examined above, this essay makes the following inference: *nuclear energy is a high-risk high-reward technology. With effective risk management – which unfortunately was lacking in Japan – the technology can enhance energy security and support economic growth and (social) development, while limiting environmental damage particularly, the emission of the notorious GHGs responsible for global warming and climate change.* From a sustainable development perspective, that is a blessing; one which renewable energy – the most preferred sustainable energy option – cannot guarantee, at least not at the moment. While renewables may be environmentally safer, they (especially wind and solar) are intermittent, seasonally variable and have low capacity factors, all of which make them incapable of solely supplying baseload electricity requirements[4] that would support economic and social wellbeing without back-up from other sources, often fossil fuel-powered (Omer, 2008; Verbruggen, 2008; Brooks et al, 2014).

It is interesting to note that in many countries with anti-nuclear sentiments, the policy responses to major nuclear disasters in the past support the inference above. Particularly, following the Fukushima and Chernobyl accidents, a number of countries assumed or reiterated existing anti-nuclear positions. For example, Italy put an end to her nuclear programme in 1987 after the Chernobyl disaster (WNA, 2014); In 2011, Japan took all her nuclear plants offline as an immediate response to Fukushima (The Guardian, 2014), while Switzerland, Germany and Belgium shut down some of their reactors and announced further plans for a complete shut down before certain future deadlines (Mez, 2012).

As at 2013 – two years after the Fukushima accident – the energy mixes in these countries were still rich in fossil fuels. In Japan, Italy, Australia and Germany fossil fuels contributed at least 65% of domestic energy production in 2013. By contrast, that figure is only 18.5% in the more nuclear-friendly France (Table 1). Only in Switzerland did renewables (including hydropower) account for more than 20% of total domestic electricity and heat energy generation. With the prospect of staying under the 2°C threshold for catastrophic global warming in 2100 looking bleak given unmitigated 2014 emissions trajectory (IPCC, 2014), it appears that what harms future generations more is not the use of nuclear energy, but the continued burning of fossil fuels where and when the option of nuclear energy is available.

Table 1: Domestic energy (electricity and heat) production by selected countries in 2013.

	Japan	Italy	Belgium	Australia	Germany	Switzerland	France
Fossil fuels (coal, oil and gas)	84.5%	69.2%	43.1%	87.4%	67.5%	7.2%	18.5%
Renewables (wind, solar, tide, geothermal and hydropower)	10%	19.3%	6.7%	11.8%	10%	44.4%	14.5%
Nuclear	0.9%	0%	35.7%	0%	8.7%	29.8%	60.8%
Biofuels, wastes and other sources	4.4%	11.5%	14.5%	12.6%	12.6%	18.5%	6.2%
Total	100%	100%	100%	100%	100%	100%	100%

Source: Generated from statistics provided by the International Energy Agency (2016). A

closer look at these nuclear phase-out decisions also shows that some of these countries allowed ample time before the proposed deadlines for the final phase-out is due (11 years in Germany, 20 in Switzerland and 24 in Belgium). Supposedly, this is to make time to develop enough renewable energy capacity to take over from fossil fuels and nuclear energy (Schreurs, 2013).

---

[4] Baseload electricity requirement is the minimum level of demand on an electrical grid over 24 hours.

Therein lies the admission that it could take some time before the reality of renewable energy unceasingly keeping the lights on, the house warm and the industries running materialises. Yet, there is every possibility that the lead time in the phase-out decisions could also be to retain the chance to reconsider their anti-nuclear positions later on, should that reality fail to materialise as and when expected. We saw such volte-face happen in Italy when the government tried, though unsuccessfully in 2011, to reverse the country's anti-nuclear stance in order to put an end to the unsustainability of energy imports. In 2008, having described the decision to end nuclear power as a "terrible mistake", the Italian government proposed to resume its generation. But, in a June 2011 referendum, the electorate firmly rejected the proposal (WNA, 2014). It is intriguing that a country with such zero tolerance for domestic nuclear energy production, imports electricity from France, Switzerland and Slovenia, whose energy mixes are nuclear-rich (Deloitte, 2015). In Germany, before Fukushima inspired a prompt recantation, the government had also scrapped initial phase-out plans, announcing in September 2010 that it would extend the life of the country's nuclear reactors by an average of 12 years (BBC, 2011). The resumption of nuclear power generation in Japan in 2013 after previous shut down (The Guardian, 2014) is yet another example. If indeed the nuclear power creates more problems than it solves, why then is its phase-out protracted or uncertain, rather than immediate and permanent, as one would reasonably expect? That the countries which made swift decisions to end nuclear energy have since resumed, or attempted to resume, its generation lends credence to the fact that despite dogged criticisms, the technology is still considered a vital component of the energy mix, for at least the near future, if energy security and lower GHG emissions are sustainable development-related goals to which we aspire.

To be clear, this essay is not oblivious of the safety risks – from nuclear plant accidents and radioactive fall-outs – that nuclear energy presents. As was cruelly demonstrated in Japan, a Fukushima-like disaster can, in no time, change the favourable outlook of nuclear energy in France or any country for that matter. But it begs the question: are current alternative sources of energy really safer? History and statistics answer in the negative. While the health consequences of the Chernobyl and Fukushima nuclear accidents have been demonised and sensationalised, history tells us that other energy sources have been equally fatal to humanity. For example, the Great Smog of London in 1952 and the 1975 Banqiao Dam Collapse in China, respectively caused by coal burning and hydropower generation and with respective direct death tolls of about 12, 000 and 26, 000 people (Xu et al., 2008, Zhang et al., 2014), arguably wrecked more damage than any nuclear plant accident ever recorded[5]. Moreover, the global average "energy deathprint", expressed in death per billion kWh of electricity generated, is lower for nuclear energy than for any of coal, oil, natural gas, biomass/biofuel, solar, wind and hydro (Table 2). Historical evidence thus suggest that nuclear energy has yet caused less harm to public health and lives than the alternatives currently available. To therefore declare the former inimical to sustainable development based on this unique unsafeness is factually inaccurate.

---

[5] Those figures, much higher when indirect mortalities from associated impacts such as diseases and famine are added, far exceed the death tolls from Chernobyl, which is adjudged to be the worst nuclear accident in history and which, according to a 2005 joint news release by WHO/IAEA/UNDP, caused about 4000 mostly cancer-related deaths in total. This essay seeks only to make a point though, not to extenuate past nuclear disasters. That would be inhumane!

Table 2: Mortality rates for each energy source in deaths per billion kWh produced.

Energy source	Mortality rate (deaths per billion kWh)
Coal – global average	100 (50% of global electricity)
Coal – China	160 (75% of China's electricity)
Coal – U.S.	15 (44% of U.S. electricity)
Oil	36 (36% of global energy, 8% of global electricity, none in U.S.)
Natural gas	4 (20% of global electricity)
Biofuel/biomass	24 (21% of global energy)
Solar (rooftop)	0.44 (<1% of global electricity)
Wind	0.15 (~1% of global electricity)
Hydro – global average	1.4 (15% of global electricity, 171,000 Banqiao dead)
Nuclear – global average	0.04 (17% of global electricity, with Chernobyl & Fukushima – none in US)

Sources: Data originally from World Health Organization; CDC; Seth Godin; and John Konrad, but table was cited from Conca and Wright (2011), and Brooks et al (2014).

The debate on nuclear proliferation is also acknowledged. The premise is that the spread of nuclear power to more countries globally could lead to spread of nuclear weapons which is detrimental to world peace and international security (Yim, 2006). However, the logical flaw in that argument is difficult to miss. While peaceful nuclear energy applications will make nuclear materials – not weapons – available to a country, nuclear materials' availability is neither a necessary nor sufficient condition for nuclear weapon acquisition. The political will/motivation and the technological capability to develop nuclear weapons are rather the key recipes (Jo and Gartzke, 2007; Sagan 2011). "If a political decision is taken, nuclear weapons can be acquired independently of any civilian nuclear power programme. In fact, historically, most countries possessing nuclear weapons acquired them before they developed peaceful applications of nuclear energy" (NEA, 2000). Besides, political measures, such as the 1970 Non-Proliferation Treaty (NEA, 2000), and technological progress, such as the introduction of proliferation-resistant Generation IV reactors (Jeong et al, 2010) are some more pragmatic measures that should be pursued to prevent weaponisation of nuclear materials originally meant for civil energy applications. If meeting the needs of current generations, while being considerate of future generations as much as possible is anything like the sustainable development we desire, then denying countries today the use of nuclear power to satisfy dire energy needs for economic and social development is simply a non-proliferation option too illogical to pursue.

V. Conclusion

Having carefully examined the cases of nuclear energy in France and Japan in the context of a world seeking to meet the needs of current generations without compromising the abilities of future generations to meet theirs, this essay concludes that nuclear energy is a crucial energy source, at least in the immediate future, while the development of renewables as efficient sustainable energy solutions continues. As far as we know today, the choice for providing stable baseload energy lies between nuclear power and fossil fuels, and as has this essay has analysed, countries that shy away from the former tend to use more of the latter.

Admittedly, careful and effective management of the technology within and between countries will be vital to any role nuclear power will play in a sustainable future. But to the extent which the technology offers humanity a reliable choice away from fossil fuels, and to the extent which it has served and can further serve as a "transition energy" between fossil fuels and renewable energy, nuclear energy is nothing but a blessing for sustainable development. For what would the world be without nuclear energy? Warmer; definitely warmer than what future generations will be able to tolerate in 2100.

REFERENCES

Abu-Khader, M. M. 2009. Recent advances in nuclear power: A review. Progress in Nuclear Energy, 51, 225-235.

Adelman, J., 2011. Fukushima Clean-up Bill $14B Over 30 Years. Available from: /http://www.bloomberg.com/news/2011-11-04/fukushima-cleanup-bill-14b-over-30-years-ministry.html (accessed 17.03.16).

Africa Energy Outlook (2014). A focus on Energy Prospects in sub-Saharan Africa. World Energy Outlook Special Report. Available at https://www.iea.org/publications/freepublications/publication/AEO_ES_English.pdf (Accessed 14.03.16)

Bachev, H. 2014. Socio-economic and environmental impacts of match 2011 Earthquake, Tsunami and Fukushima nuclear accident in Japan. *Journal of Environmental Management and Tourism,* 5, 127-222.

BBC 2011. Germany: Nuclear power plants to close by 2022. http://www.bbc.co.uk/news/world-europe-13592208 (Accessed 30.03.16)

Blowers, A. 1999. Nuclear waste and landscapes of risk. Landscape Research, 24, 241-264.

Bouzarovski, S. 2014. Energy poverty in the European Union: Landscapes of vulnerability. *Wiley Interdisciplinary Reviews: Energy and Environment,* 3, 276-289.

Brook, B. W., Alonso, A., Meneley, D. A., Misak, J., Blees, T. & Van Erp, J. B. 2014. Why nuclear energy is sustainable and has to be part of the energy mix. *Sustainable Materials and Technologies,* 1-2, 8-16.

Bruntland Commission. 1987. Our Common Future: Report of the World Commission on Environment and Development.

Butler, D. 2010. France digs deep for nuclear waste. *Nature,* 466, 804-805.

Conca. 2012. How Deadly Is Your Kilowatt? We Rank The Killer Energy Sources. http://www.forbes.com/sites/jamesconca/2012/06/10/energys-deathprint-a-price-always-paid/#7e94ecf849d2 (Accessed 27.03.2016)

Conca J.L. and Wright J. 2011. The Cost of Energy – Ethics and Economics. Waste Management 2010, Phoenix, AZ, paper 10494, 1 – 13

Dautray, R., Friedel, J. & Bréchet, Y. 2012. Nuclear energy in France today and tomorrow: Second to Fourth generations. Comptes Rendus Physique, 13, 480-518.

Deloitte 2015. European Energy Market Reform. Country Profile: Italy. https://www2.deloitte.com/content/dam/Deloitte/global/Documents/Energy-and-Resources/gx-er-market-reform-italy.pdf (Accessed 18.03.16)

EDF Electricité de France History (nd). http://www.fundinguniverse.com/company-histories/electricité-de-france-history/ (Accessed 15.03.16)

Gallardo, A. H. & Marui, A. 2016. The aftermath of the Fukushima nuclear accident: Measures to contain groundwater contamination. *Science of The Total Environment,* 547, 261-268.

Hyde, R., Ishikawa, M., Myhrvold, N., Nuckolls, J. & Wood, L. 2008. Nuclear fission power for 21st century needs: Enabling technologies for large-scale, low-risk, affordable nuclear electricity. *Progress in Nuclear Energy,* 50, 82-91.

IAEA, International Atomic Energy Agency (nd). Nuclear Accidents – and how they are ranked. https://docs.google.com/spreadsheets/d/1j7_05tUdnUCrHCYTplWKWavHjUW6zcIw3 WlgyLf5syA/edit?hl=en#gid=1 (Accessed 15.03.16)

IEA, International Energy Agency. 2006. IEA Statistics by Country/Region, France, /http://www.iea.org/ (Accessed 16.03.16)

IEA, International Energy Agency. 2015. Key world energy statistics. Available at https://www.iea.org/publications/freepublications/publication/KeyWorld_Statistics_2015 .pdf (Accessed 16.03.16)

IEA, International Energy Agency. 2016. Statistics by country http://www.iea.org/statistics/statisticssearch/ (Accessed 20.03.16)

IPCC, 2014. Summary for Policymakers. In: Climate Change 2014: Mitigation of Climate Change. Contribution of Working Group III to the Fifth Assessment Report of the Intergovernmental Panel on Climate Change. Cambridge University Press, Cambridge, United Kingdom and New York, NY, USA

Jeong, H.Y., Kim, Y.I., Lee, Y.B., Ha, K.S., Won, B.C., Lee, D.U. & Hahn, D. 2010. A 'must-go path' scenario for sustainable development and the role of nuclear energy in the 21st century. *Energy Policy,* 38, 1962-1968.

Jo, D. J. & Gartzke, E. 2007. Determinants of nuclear weapons proliferation. Journal of Conflict Resolution, 51, 167-194.

Kessides, I. N. 2012. The future of the nuclear industry reconsidered: Risks, uncertainties, and continued promise. *Energy Policy,* 48, 185-208.

Lee, R. P. & Gloaguen, S. 2015. Path-dependence, lock-in, and student perceptions of nuclear energy in France: Implications from a pilot study. *Energy Research and Social Science,* 8, 86-99.
Lee, C.C. & Chiu, Y.B. 2011. Nuclear energy consumption, oil prices, and economic growth: Evidence from highly industrialized countries. *Energy Economics,* 33, 236-248.

Mbarek, M. B., Khairallah, R. & Feki, R. 2015. Causality relationships between renewable energy, nuclear energy and economic growth in France. *Environment Systems and Decisions,* 35, 133-142.

Mez, L. 2012. Nuclear energy–Any solution for sustainability and climate protection? Energy Policy, 48, 56-63.

Ming, Z., Yingxin, L., Shaojie, O., Hui, S. & Chunxue, L. 2016. Nuclear energy in the Post-Fukushima Era: Research on the developments of the Chinese and worldwide nuclear power industries. *Renewable and Sustainable Energy Reviews,* 58, 147-156.

Naser, H. 2015. Can nuclear energy stimulates economic growth? Evidence from highly industrialised countries. *International Journal of Energy Economics and Policy,* 5, 164-173.

NEA, Nuclear Energy Agency 2000. Nuclear Energy in a Sustainable Development Perspective. Organisation for Economic Co-Operation and Development. Available at https://www.oecd-nea.org/ndd/docs/2000/nddsustdev.pdf (Accessed 15.03.16)

Omer, A. M. 2008. Energy, environment and sustainable development. Renewable and Sustainable Energy Reviews, 12, 2265-2300

Petit, P. 2013. France and Germany nuclear energy policies revisited: A veblenian appraisal. Panoeconomicus, 60, 687-698.

Reuters, 2011. Japan sees atomic power cost up by at least 50 pct by 2030. Available from: /http://www.reuters.com/article/2011/12/06/japan-nuclear-cost-idUSL3E7N60MR20111206S (Accessed 17.03.16).

Sagan, S. D. 2011. The causes of nuclear weapons proliferation. Annual Review of Political Science. 14, 225-244.

Shibata, H. 1983. The energy crises and Japanese response. *Resources and Energy,* 5, 129-154.

Schreurs M. A. 2013. Orchestrating a Low-Carbon Energy Revolution Without Nuclear: Germany's Response to the Fukushima Nuclear Crisis. Theoretical Inquiries in Law, 14, 83 - 108.

Srinivasan, T. N. & Gopi Rethinaraj, T. S. 2013. Fukushima and thereafter: Reassessment of risks of nuclear power. *Energy Policy,* 52, 726-736.

Stainer, A. & Stainer, L. 1995. Young people's risk perception of nuclear power - a European viewpoint. *International Journal of Global Energy Issues,* 7, (d) 261-270.

Statistica. 2015. The Statistics Portal. Statistics and Studies from more than 18 000 sources. Available at http://www.statista.com/statistics/263492/electricity-prices-in-selected-countries/ (Accessed 19.03.16)

Teräväinen, T., Lehtonen, M. & Martiskainen, M. 2011. Climate change, energy security, and risk—debating nuclear new build in Finland, France and the UK. *Energy Policy,* 39, 3434-3442.

The Economist 2011. Nuclear Power. When the Steam clears. Available at http://www.economist.com/node/18441163 (Accessed 18.03.16)

The Guardian 2014. First Japanese nuclear power plant since Fukushima to resume operations. http://www.theguardian.com/world/2014/nov/07/japanese-nuclear-power-plant-fukushima-restart (Accessed 27.03.16)

The National Diet of Japan. 2012. The official report of The Fukushima Nuclear Accident Independent Investigation Commission. Available at http://ieer.org/wp/wp-content/uploads/2012/07/Fukushima_NAIIC_report_lo_res3.pdf (Accessed 17.03.16)

Tokimatsu, K., Fujino, J. I., Konishi, S., Ogawa, Y. & Yamaji, K. 2003. Role of nuclear fusion in future energy systems and the environment under future uncertainties. Energy Policy, 31, 775-797.

Verbruggen, A. 2008. Renewable and nuclear power: A common future? Energy Policy, 36, 4036-4047.

WHO/IAEA/UNDP 2005. Chernobyl: the true scale of the accident. http://www.who.int/mediacentre/news/releases/2005/pr38/en/ (Accessed. 31.03.16)

Wiegman, O., Gutteling, J. M. & Cadet, B. 1995. Perception of nuclear energy and coal in France and the Netherlands. *Risk Analysis,* 15, 513-521.

World Bank 2016. http://data.worldbank.org/indicator/NY.GDP.TOTL.RT.ZS (Accessed 16.03.16).

WNA, World Nuclear Association 2013. National policies. Radioactive Waste Management. http://www.world-nuclear.org/information-library/nuclear-fuel-cycle/nuclear-wastes/appendices/radioactive-waste-management-appendix-3-national-p.aspx (Accessed 16.03.16).

WNA, World Nuclear Association 2014. Nuclear Power in Italy. http://www.world-nuclear.org/information-library/country-profiles/countries-g-n/italy.aspx (Accessed 30.03.16)

WNA, World Nuclear Association 2015. Nuclear Power in France. http://www.world-nuclear.org/information-library/country-profiles/countries-a-f/france.aspx (Accessed 16.03.16).

WNA, World Nuclear Association 2016. Nuclear Power in Japan. http://www.world-nuclear.org/information-library/country-profiles/countries-g-n/japan-nuclear-power.aspx (Accessed 16.03.16).

Xu, Y., Zhang, L. & Jia, J. 2008. Lessons from catastrophic dam failures in August 1975 in Zhumadian, China. Geotechnical Special Publication, 2008. 162-169.

Yim, M.S. 2006. Nuclear non-proliferation and the future expansion of nuclear power. Progress in Nuclear Energy, 48, 504-524.

Zhang, D., Liu, J. & Li, B. 2014. Tackling air pollution in China-What do we learn from the great smog of 1950s in London. Sustainability (Switzerland), 6, 5322-5338.